FUTURE SCOUTING®

DAMIEN LUTZ

Copyright © Damien Lutz.

All rights reserved. No part of this publication may be reproduced, distributed, or transmitted in any form or by any means, including photocopying, recording, or other electronic or mechanical methods, without the prior written permission of the publisher, except in the case of brief quotations embodied in critical reviews and certain other non-commercial uses permitted by copyright law. For permission requests, write to the publisher, addressed "Attention: Permissions Coordinator," at the address below.

ISBN: 978-0-9946275-7-5 (Paperback)
ISBN: 978-0-9946275-8-2 (eBook)

Cover design by Damien Lutz.
First edition 2021.
Second edition 2021.
Third edition 2022.

futurescouting.com.au

Table of Contents

Table of Contents iii
Figures & Tables iv
Acknowledgment v
1 Introduction 1
2 Speculative Design 5
3 Speculative Design Methods & Tools 25
4 Future Scouting 37
5 Prepare to time travel 45
6 Catch a signal 53
7 Hunt an invention 73
8 Release your invention 89
9 Hero and future world 97
10 Prototype and share 115
11 Home 133
Appendix A Prototype Examples 137
Appendix B Notes 147
About the author 151

Figures & Tables

Figure 1.0 - Future types and potentials 6

Figure 2.0 - Landscape of design practices 10

Figure 3.0 - Altering future direction with speculative design 20

Figure 4.0 - The Future Scouting method 38

Figure 5.0 - Worksheet: Future Scouting Dashboard .. 46

Figure 6.0 - Worksheet: Signal Tracer 58

Figure 7.0 - Worksheet: Research Map 74

Figure 8.0 - Worksheet: Scenario Futures Wheel 90

Figure 9.0 - Artefact: Future Persona 104

Figure 9.1 - Worksheet: Story Futures Wheel 108

Figure 9.2 - Artefact: Location Snapshot 110

Figure 10.0 - Table: Prototype mediums 121

Acknowledgment

Future Scouting® draws on the great work of Future Studies and speculative design practitioners past and present, such as Anthony Dunne, Fiona Raby, Stuart Candy, Scott Smith, Masaki Iwabuchi, David Brian Johnson, Elliott P. Montgomery, Frog Design, SpeculativeEdu.com, and many others, who have in various ways expanded the idea of what design can be. I'm very thankful for the inspiration their work has given me, and I hope this book flows that inspiration forward to many others.

1
Introduction

Congratulations—you hold in your hand a multi-future time machine!

By the time you finish this book, you'll be picking up signals of emerging trends and plucking inventions from one of the many tomorrows—and you won't need expert forecasting or complex data analysis skills to do it.

Future Scouting is a fun and practical way to use science fiction world-building techniques to create fantastic future inventions to inspire others to think about how they really want the future to be.

On your journeys into the myriad of futures, you'll be tempted to visit just the utopias where all your dreams come true. But everyone's dreams are different, and striving for one utopia might destroy all others, and possibly all future itself.

But while the promising finds of utopia can have a more immediate positive impact on an audience, only dystopian bounties shed light on how to prevent dark trends enabling the end of the world.

It is only when we consider the full spectrum of human behaviour that we see a more diverse and inclusive tomorrow. From these visions, we can begin to map out the more realistic protopian [1] versions of tomorrow—not perfect worlds, but always improving.

But only storytellers explore these darker futures, while designers strive to remain ethical by avoiding dark patterns.

Why should only storytellers get to play the villains?

Future Scouting combines storytelling with design to empower designers to explore dystopias as much as the utopias while maintaining a focus on core values.

If you dare tread this perilous yet exciting path, let me introduce you to the mind-bending practice of Speculative Design—stealing a dystopian artefact from a future that may never exist to prevent it from plaguing the world tomorrow.

There's a dangerous paradox in there, but let's keep moving.

Be warned, however, for the adventures of a Future Scout comes with great responsibility—your finds might cause the disassembling of something a fellow designer creates in the near future, or perhaps it will be something created by your older self!

There are many articles, websites, and blogs sharing wonderful insights about speculative design, but they are scattered far and wide across the internet, while many discussions are heavily academic in language.

Saving the world should be fun.

Based on methods I've used to design future technology and worlds for science fiction novels—and drawing on the great work of Future Studies and speculative design practitioners past and present—*Future Scouting* combines product design, design thinking, life-centred design, and science fiction world-building into a speculative design method that is fun and accessible. With supporting resources downloadable from futurescouting.com.au, *Future Scouting* eases the discovery process of speculative design to encourage exploration and experimentation by summarising its purpose, tools, and methods, before flinging you into the future with a step-by-step process to get you saving the world and back to the present well before the next pandemic.

So, if you've ever imagined a better version of your favourite device, a new social system that considers the environment, or innovative ways to improve the lives of others, then let *Future Scouting* show you how to design many future possibilities and explore their impact on tomorrow.

Calling all Sarah Connors dreaming of better futures— good luck!

2
Speculative Design

In a time when an invisible bug shuts down the world, bushfires turn the skies Bladerunner orange, and the truth becomes fake news while fake news becomes truth, it's undeniable that science fiction can become science fact.

Yet the orange skies remain empty of hoverboards, jetpacks, flying cars, and so many other child-hood dreams.

You wouldn't be blamed for thinking we have ended up in a dystopia.

But when you look a little deeper into the fringes of the internet, media, and pop culture, interesting experimentations flicker into view—a citizen-designed carbon economy; a euthanasia smartwatch; artificially designed fruits; and the ability of female human bodies to reproduce endangered species.

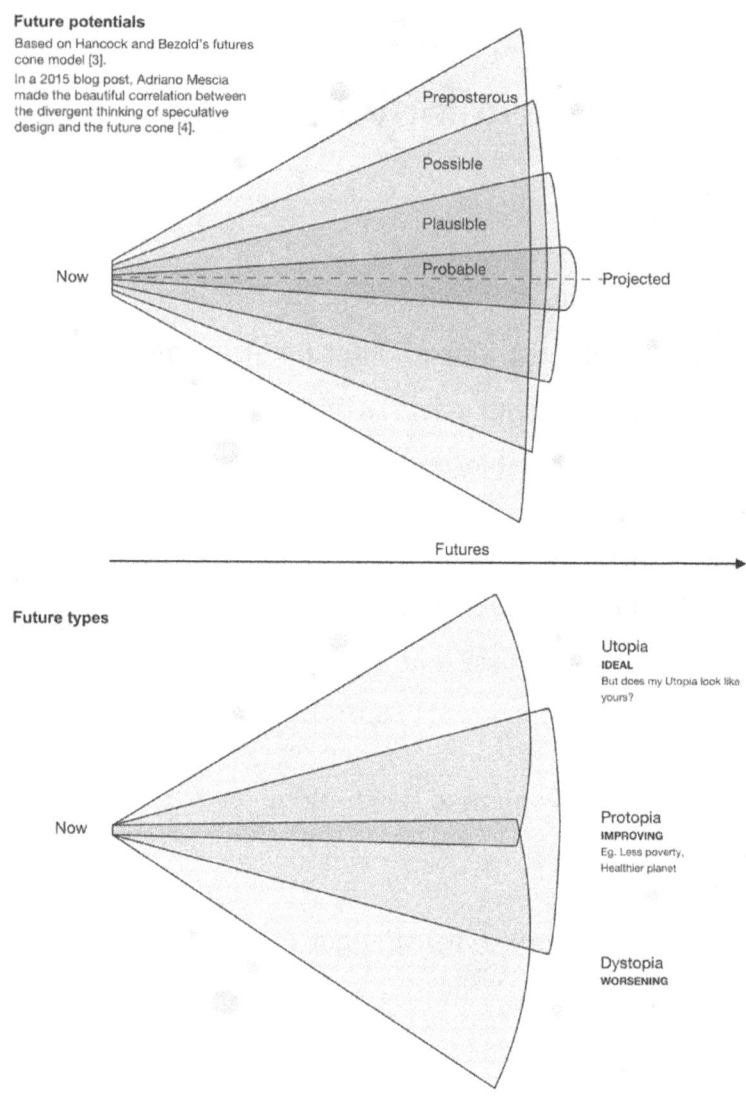

Figure 1.0 - Future types and potentials

While these concepts may not appeal to general ideas of utopia, they are potential seeds of *protopias*—future states where things are not utopian perfect, but always improving, freeing us from the limiting duality of utopia versus dystopia, and therefore more realistic to hope for.

These experiments are products of speculative design—an evolving design practice that encourages the world to think further ahead and wider in possibility.

Speculative design combines design thinking methods with the story-telling and future-world-building techniques of science fiction. Also referred to as 'design fiction'—and incorporating other futures studies practices such as critical design—speculative design's purpose is to create prototypes of imagined future experiences to generate wider discussion about what is possible by 'collectively redefining our relationship to reality' [1].

The prototyped artefacts produced and shared back encourage audiences to think beyond the probable future, sparking their subconsciouses to include scenarios from the realms of the plausible and potential, from dystopia, protopia, and utopia (Figure 1.0).

Speculative design produces questions—it is not predicting or problem-solving. Unlike the data-informed systematic inquiry of foresight and future research, which attempt to clarify possible and probable futures, speculative design is 'more intuitive, less disciplined, and less data-oriented' [2], and therefore takes the design process beyond the expected and commercial (Figure 2.0). The prototypes are not required to be fit for practical application (although they can be)—their purpose is to generate reframing of our perspectives of the future to reveal to us unseen alternate trajectories of cultural, technological, environmental, economic, and socio-political trends.

These fantastic artefacts of fact and science fiction may come in the form of a physical or digital product, video, documentary, book, manual, website, sculpture, etc.

Science fiction casts present and imagined trajectories into the future to warn and inspire us so we might change the direction in which we are headed. Designing for science fiction scenarios can have a similar effect on a designer's thinking as it allows them to do what they are taught not to do—explore dark patterns and trends—to produce something ironic and provocative that by its very design tells a story about the future world it belongs to. This playful projection of what future worlds might exist, how we may have arrived there, and how people behave in them, expands a designer's peripheral creativity and deepens their awareness of human behaviour and psychology.

Designing for the future is as much about what we learn about ourselves during the process as sparking braver thinking and behaviour in others from the end product.

Speculative Design is fun, exciting, and humbling.

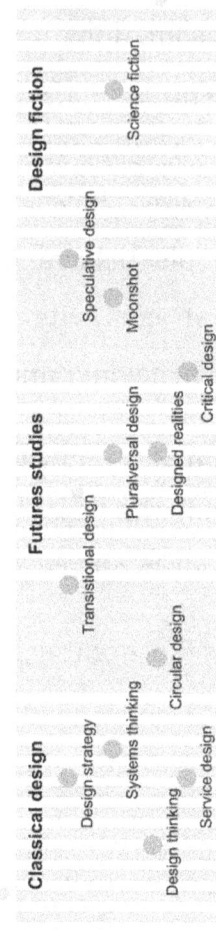

Figure 2.0 - Landscape of design practices

Prototype Examples

Speculative Design prototypes are 'diegetic' artefacts, meaning they are narrative-based, suggesting by their form and function the nature of their imagined future world while still leaving room for the audience to fill in the gaps with their imagination.

This is similar to writing speculative and science fiction, where the author conveys essential information of their imagined world without info-dumping extraneous detail, allowing readers' imagination to concoct their own version of the future. These personal interpretations encourage participants to articulate both their preferred future and their reasoning.

Museum of Futures

Not to be confused with The Museum of the Future under construction in Dubai, the Museum of Futures is an interactive virtual gallery created by Sydney-based creatives Claire Marshall and Mel Rumble. As a virtual visitor, you can tour the online gallery to explore artefacts from alternate futures created during experiential workshops facilitated by Marshall's creative company *iflabs*.

The latest exhibition, *Pandemic Pivots*, explores two possible futures—a utopia where the world tackles climate change, and a dystopia where extreme climate change has altered our way of life forever. Previous exhibitions explored the Future of Work, the Future of Australia, and The Future of Food. The virtual Museum of Futures showcases the fascinating artefacts from all exhibitions, including a 'repurposable' toy made in 2035; a sculpture commemorating the 2030 introduction of The Native Species Title Act that 'gave animals rights to all native forests and bushlands across Australia'; and a bottle of Australian air from the early 2020s'.

museumoffutures.com

Extrapolation Factory

Founded by creatives Chris Woebken and Elliott P. Montgomery, Extrapolation Factory use their own experimental, collaborative methods to explore democratised futures by creating hypothetical future props. They often embed their provocative props into real-world scenarios.

'*99¢ Futures*', the Factory's first project—and my personal favourite—began with a selection of possible future scenarios expanded into small stories. Related product ideas generated from the stories ('Benzene Vapor Refills', 'Mars Survival Kits', and 'Triple-Nipple Baby Bottles') were created via rapid-prototyping and then stocked for purchase in a real-world 99¢ store amongst usual products. This guerrilla installation of a 99¢ store with a Time-Warp sale provoked conversation between strangers about the future scenarios triggered by these specials.

extrapolationfactory.com/99-Futures

I wanna deliver a shark...

Addressing the potential future issues of overpopulation and species extinction, Japanese designer Ai Hasegawa proposed future biomedical technology that enables women to use their reproductive system to birth endangered species... that they could also eat. Hasegawa likes to eat dolphin but recognises this adds to the animal's endangerment. She also wants to not waste her reproductive system but doesn't want children. Hasegawa took her concept further by conducting a scientific study to realise a scheme of how this could actually look.

aihasegawa.info/i-wanna-deliver-a-shark

Euthanasia for everyone

The Speculative Design Provocative Award goes to *Soulaje*, a self-administered euthanasia wearable proposing to give people control over the place and time of their death. Consisting of a vial of some death-inducing potion attached to a smartwatch, you simply tap and die. You can't get much more user-friendly than that! Created by Design Friction studio, All-Party Parliamentary Design and Innovation Group (APDIG), and Age UK, the Soulaje concept arose from exploring loneliness for the ageing.

You can find these and more examples in Appendix B.

Founded in values

Speculative Design took form in the early 1990s as a response by designers to the 1980's embrace of capitalism. Designers were frustrated by serving a mass-market machine without being allowed a thought beyond form and function, and they began to question their role in consumerism's impact on the planet. Founded in the challenge to dominant forces, speculative design remained a curio or unheard of to the wider design folk. But with the mass sharing afforded by the internet, word is getting out that designers are allowed to have values, and that ignoring them might be enabling a future that opposes them.

Why design for science fiction

1. Expand your idea of what's possible

By projecting design thinking on a fictitious future scenario—a scenario you choose—*you'll travel to the future*.

While the time travel of speculation is done in thought only, this powerful exercise will enlighten you to greater possibilities, such as what emerging technologies to design for, what current 'good' and 'bad' user trends might prevail, and how dark patterns are employed.

When you come back, you'll be a little more aware of how your designs might be hindering or enabling these trends.

And that means thinking about ethics and evil.

2. Play Evil, Learn Ethics

As ethical designers, we avoid dark patterns that exploit users' base desires and behaviours.

But experimenting with flawed technology is a deliberate sci-fi world-building tactic to explore the repercussions of following current paths.

By playing the provocateur, you'll learn more about humanity by exploring people's faults as well as their dreams, and this can provide previously unseen possibilities on how to transcend both. And when you see the trajectories of your own current thinking, you'll be able to make design decisions more aligned with your values and perhaps architect a future world more aligned with these.

Exploring the bad as well as the good in humanity forces you to travel outside your comfort zone, and this is great practice for tapping into diversity, emotion, drama, and irony to enhance the inclusivity and personality of any other design work you do.

3. Get a self-worth boost

Designing for fiction allows you to apply your skills and passion to something unrestrained by legacy technologies, tightening budget restrictions, or conflicting stakeholder needs.

When you practice your design on a need or problem of your own choosing—and on your own terms—you'll be delightfully surprised to see how much you've grown and how you've personalised your own design process. This is pure joy in of itself and a celebration every designer deserves.

4. Save the world

Much of the planet's design thinking is enslaved to the consumer machine, encouraging people to buy stuff and do things that harm the planet, hurt our animal friends, and exploit people in poverty.

Scientists warn us that the ugly is going to get a lot uglier. But the good news is that designers have the superpowers to improve things—if we unplug some of our design energy from the consumer machine and refocus it.

This means we need to invest our own time—get out of our work-from-home clothes for an hour or two a week, get our superhero spandex on, and choose a real-world problem that is close to our hearts. Then design the hell out of it. Spin it into the future with speculative design and bring back new perspectives to change direction (Figure 3.0).

We might not fully solve problems with utopian solutions, but we might improve them into new protopian states.

Imagine if every designer focused just a little of their own time on speculating future resolutions and then shared their projects with others…surely the generated discussions would enrich social, scientific, technological, and environmental progress.

5. Apply future thinking in business

At the time of writing, a number of well-respected businesses are already utilising speculative design:

- Innovation company SciFutures "accelerates innovation with sci-fi prototyping"
- The creative studio Imagination of Things uses design and fiction to craft meaningful stories for their clients
- UP Future Sight uses design fiction, strategic planning, and trend forecasting to help companies plan ahead
- Rito Services hosts immersive experiences that blend fiction and reality to facilitate views about the future
- Envisioning maps fictional technologies and connects them to their sci-fact counterparts

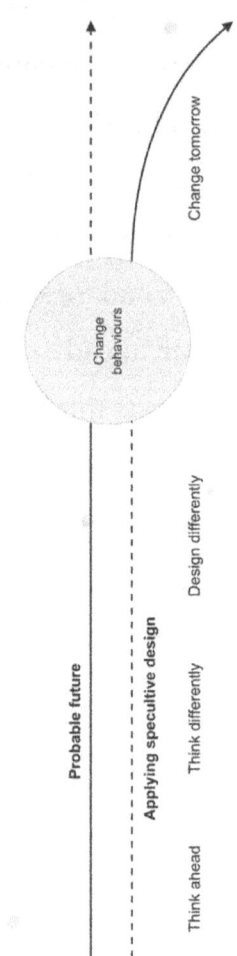

Figure 3.0 - Altering future direction with speculative design

The application of speculative design in business includes:

- Providing an innovative starting point for companies seeking new markets and/or to future-proof their existing products
- Governments and their partner agencies aiming to innovate policy making
- Designers creating artifices that function as design probes into their own processes to make them more future-aware
- Designers and artists creating thought pieces to awaken the audience's crisis-numbed, news-saturated, consumer-overloaded minds so they may articulate their preferred futures

6. Most importantly, have some well-earned fun with design

Earlier this year, I got to explore designing a speculative concept for sustaining peace for the *Futuring Peace* design competition.

Speaking from that experience, I can tell you that playing with your craft without the usual project restraints induces wonder and artistry and reconnects you with your passion for design. Remember how exciting it was when you first started playing with web design when you didn't know you couldn't place anything anywhere on a web page? ('Why is everything in boring tables? I have to consider code??')

Designing for the future allows you to reapply that same limitless imagination in a more thoughtful, constructive way.

7. And when you're done, add a new dimension to your folio

For the more pragmatic-minded, you'll be happy to hear that designing for science fiction also creates something tangible to add to your folio.

Showcase how you experiment with design beyond your job and how you applied design thinking by packaging up your fantastic future design into a case study, future promotional web page, or blog article.

You'll show you're a design explorer who thinks ahead and scouts the fringes of innovation.

Sharing your process, and choice of future scenario will provide an exciting and personal insight into who you really are as a designer, hopefully making you stand out to an employer or business partner more aligned with your values and vision.

Testing the futures

Considering the number of potential futures and future types awaiting configuration by today's choices, how could we know which future we really need without testing their *experience*? Designing for the future is about making mistakes in the virtual realm of fiction so we can avoid them in reality. And sometimes, the process plucks a jewel of pure genius from the ether and shifts the world into a new and better direction—like X's *Loon* project "expanding internet connectivity with stratospheric balloons [6]". While *Loon* shut down early 2021, its moonshot work has planted shiny silver seeds for future innovation.

This is what Speculative Design does, it brings the infinite possibilities to us like servants with a platter, so we might realise we are not the spoilt consumer royalty that greedy conglomerates would have us believe, but highly tuned, empathetic individuals imbued with the power of future-shaping choice.

We just need the right tools.

3
Speculative Design Methods & Tools

While your time machine is still warming up, let's look at the speculative design devices that inspired *Future Scouting*.

The methods and tools below can be used on their own or as part of a larger and more serious design project, such as future proofing a company's roadmap.

How they are used depends on the purpose of the project—exploring future business opportunities, generating discussion on a particular subject, or designing for fiction or art. The importance of the outcome also dictates the approach—planning a company's next product line requires rigorous data analysis and synthesis, as well as generating provocative ideas from an ideation workshop.

Speculative design done as a purely provocative project (what this book teaches) starts with speculating ideas from signals, creating a narrative around them to produce prototypes, and then shares these with an audience to provoke discussion.

Strategising a company's direction starts with more researched-informed signal tracing, uses speculative design to produce alternate futures and/or future artefacts, reviews and optimizes a preferred future, then back-casts a step-by-step strategy based on what changes need to happen to make the preferred future become realised.

Speculative design tools

Designing for future scenarios is a very creative and imaginative process, so designers should feel free to draw on tools and techniques from both design thinking and the creative arts—storyboards, role-plays, games, interviews and questionnaires, writing, animation, film, and so on.

Here are a few tools developed by others to help you ideate your futures.

The Tarot Cards of Tech

The Tarot Cards of Tech [1] inspires product designers to envision alternate futures of their product or idea. From the unseen and negative to the hoped-for and ideal, the process helps creators slow down and consider more deeply the impact of their product beyond the obvious needs of its users.

Created by *Artefact*, a U.S. strategy and design firm.

Future scan

Another great tool for inspiring ideation is the beautiful *Future Scan* map [2] from the *Board of Innovation*. The map offers over 150 potential future scenarios based on next-gen technologies and societal changes. It can be used either as a simple exercise by choosing a number from the map and brainstorming that scenario, or as one tool in a larger speculative design process.

Future Timeline

The Future Timeline [3] is a fascinating, resource-packed website collating predictions for the next two centuries and beyond, including data and technology trends, articles on emerging technology, and much more. Created by London-based writer and futurist, *William James Fox*, and updated regularly since 2008, *The Future Timeline* began as a small and quirky website showcasing a brief list of future predictions. Insightful and provocative, the idea attracted many fans and contributors over the years, growing into a vast, community-based project for futurology enthusiasts from all over the world.

Use the *Future Timeline* as a resource for speculative design projects to generate ideas about the year or technology you are designing for.

Cover story

Cover Story [4] is an imagination game for businesses to brainstorm their ultimate future state by designing a magazine cover announcing their success. Players are asked to disregard existing limitations to envision a "best-case scenario for their company" and ideate the magazine cover in terms of headline, quotes, images, etc. The process can reveal previously unimagined ideas and directions for the company to explore.

Based on *The Cover Story Vision® Canvas* [5] created by *David Sibbet*, of *Grove International*.

The Thing From The Future

The Thing From The Future [6] is an award-winning ideation card game for individuals or teams that challenges players to articulate objects from alternative futures. The deck consists of 108 cards of four suits—the type of future the thing comes from, its thematic context, its basic form, and the emotion the object generates in someone from the present.

Players take turns placing cards onto the table until they have a card from each suit—this forms the creative prompt players must ideate on. The players then vote on which concept is the best, funniest, most ironic, etc. The winner of each round keeps the cards played, and whoever has the most cards when the game ends is the winner. The many speculative designs produced can be used for further exploration and innovation.

Created by foresight practitioner and educator *Stuart Candy* and media and games educator *Jeff Watson*.

Frog Design's 'headline' method

Frog Design uses its headline method [7] to help businesses imagine a future 5–15 years ahead and the products and services that might be needed. Their process focuses on creating future news headlines:

- **Step 1:** Define the timeframe
- **Step 2:** Research potential future risks to the business
- **Step 3:** Ideate a future world
- **Step 4:** Create many hypothetical news headlines for the imagined future
- **Step 5:** Choose key headlines and workshop how the business might address these issues

Ethnographic Experiential Futures

Also created by *Stuart Candy*, and design researcher/futurist *Kelly Kornet*, *Ethnographic Experiential Futures* [8] combines ethnographic futures research—a recording and surfacing of existing concepts of the future—with experiential futures—a future ideation method that uses multimedia, multi-sensory, and other story-telling techniques—to create new future concepts, with a focus on stimulating the senses. This wonderful approach makes speculative design prototypes more of an experience with more impact and relatability for the discussion phase.

Candy and Kornet suggest the following process:

- **Map:** Research how people currently imagine the future to reveal what is considered possible, probable, and/or preferred
- **Multiply:** Ideate alternative scenarios by challenging the restraints applied to the above
- **Mediate:** Develop these concepts into tangible, interactive experiences
- **Mount:** Share via exhibition, interactive workshop, 'guerrilla future' installation, or other means
- **Map:** Record response and feedback

Science fiction prototyping

Futurist *Brian David Johnson* developed science fiction prototyping [9] in 2010 in response to the business challenge of anticipating market needs for products after they reach the end of their design and production cycle. The method involves generating a three-act story as a speculative design prototype to innovate within the areas of science, engineering, business and socio-politics.

- **Act 1:** Ideate a future—choose a technology to explore and ideate a future incarnation and its experience
- **Act 2:** Scientific inflection point and ramifications—introduce a catastrophe and story its repercussions on the user
- **Act 3:** Human inflection point—explore and articulate how a user would remedy their situation

The story is then used for reflection discussions and further product innovation.

Speculative design methods

Provocation

Used on its own, a speculative design process starts in the future and 'casts' the experiment back to the present.

- **Step 1**—Identify 'signals' of emerging technologies and trends
  o As speculative design deals with less expected futures, the process begins with identifying more fringe technologies and culture trends. Look for these 'weak signals' in emerging technologies and experimental uses

- **Step 2**—Ideate a future product
  o Drawing on Step 1, use various speculative design tools to ideate your future product (See Speculative design tools below)

- **Step 4**—Storify the product
  o Design a diegetic artefact that might solve a problem in your future world. Make sure the artefact's form and function tell a story about its future world and inspires the audience's imagination to fill in the gaps. These personal interpretations infuse the discussion in the next step with energy and personality

- **Step 5**—Share to generate discussion
    - Whoever your audience is—stakeholders, fellow designers, social media followers, or the general public—make it easy for them to offer their interpretation of your product. Be prepared with questions that encourage them to articulate why they do or don't like your future concept, how it might affect them personally, and what they might prefer.

Strategy

When speculative design is part of a larger problem-solving project, the process might consist of the following:

- **Step 1**—Ideate a future
    - Choose a time frame - If you're looking to innovate for a business, Frog Design's headline method suggests casting 5 to 15 years ahead, depending on the industry. For an industry that updates its product line every 3 years, then designing for at least 3 years in the future makes sense

and designing for 9 years would be akin to looking 3 cycles ahead. For industries with slower growth, and companies leading in their industry, looking even further into the future can help with long-term strategy and spotting new threats. But the more distant the future you aim for, the more the landscape blurs, and the casting becomes less viable.

  o Ask "What if?" - Think in terms of potential growth areas, risks, existing related technologies, emerging and proposed new technologies, competitors, user needs, behaviour trends, supply chains and related industries. Research these areas and map out where they might be in your chosen timeframe, and how they might affect each other. Explore both best- and worst-case scenarios.

- **Step 2**—Ideate future needs and problems
  o Ideate user personas inhabiting the future you devise to identify their needs and

explore future business problems to be solved.

- **Step 3**—Design a solution and 'storify'
- **Step 4**—Share and record
- **Step 5**—Iterate and innovate
  - Summarise the responses and findings and iterate new versions to begin a follow-on product innovation process.

To put into perspective how all these methods and tools might work together, let me introduce you to the *Future Scouting* method.

4
Future Scouting

Speculative Design isn't just about exploring the future, it's also about experimenting with the process. Once you get familiar with the approach, you can mix the tools and methods and invent your own processes. *Future Scouting* is one method that utilises various tools from multiple practices to create an introduction to speculative design.

When designing for today, design thinking often starts with divergent research, but any innovations are then shaped by the commercial need to validate people's desirability, technical feasibility and business viability.

Future Scouting starts by ideating a future invention first and then extrudes people and future landscapes from these inventions.

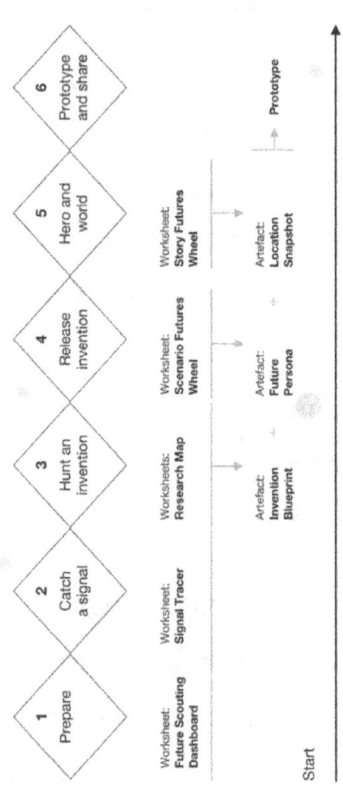

Figure 4.0 - The Future Scouting method

By working in reverse from an idea that's allowed to evolve unrestrained, speculative design unleashes design thinking from commercial safety and takes design—and the audience—into realms beyond the probable where joys of unexpected delight may otherwise be hidden from us.

The goal

Future Scouting's ultimate goal is the same as that of speculative design—to improve our current probable future by empowering others with a broader thinking about the future [1]. *Future Scouting* generates content and artefacts to share with your audience. You may not use all of them, and some of the method's exercises may purely be a bridge to the next, but you most likely won't know how you'll combine your artefacts until you're finished.

The principles

As mentioned in Chapter 1, *Future Scouting* draws on the great Future Studies and speculative design work of devoted practitioners such as Dunne & Raby, Stuart Candy, Scott Smith, Masaki Iwabuchi, David Brian Johnson, Frog Design, SpeculativeEdu, and others, to combine product design, design thinking, life-centred design, and science fiction world-building into a speculative design method that is fun, value-driven, and accessible.

Realism

While audiences love science fiction for its escapism, concepts based in realism help them suspend their belief and immerse themselves in the proposed fantasy. Future Scouting conceives future concepts from research to ground the concept in realism.

Core values

To maintain the spirit of speculative design's origins, *Future Scouting* grounds itself in designing according to values. However, the *Future Scouting* process can also expose unintended, black-mirror type consequences of best intentions to design with a focus on one value.

Accessibility

No complex data analysis or forecasting skills are required.

The steps

As you journey through the future, you'll collect artefacts you create to form a final prototype to share back with today's audience.

Step 1: Prepare
Worksheet: Future Scouting Dashboard

- Centre yourself, know your purpose—choose a Key Value to champion
- Know your assumptions of the future
- Set your range

Step 2: Catch a signal
Worksheet: Signal Tracer

- Identify an emerging phenomenon

Step 3: Hunt an invention

Worksheet: Research Map
Artefact: Invention Blueprint

- Research the signal to develop your future idea

Step 4: Release your invention
Worksheet: Scenario Futures Wheel

- Map your invention's impact

Step 5: Hero and future world
Worksheet: Story Futures Wheel
Artefacts: Future Persona & Location Snapshot

- Meet the future world and its inhabitants

Step 6: Prototype and share

- Create and share your prototype

The resources

Each step uses a 'worksheet' and/or 'artefact' designed specifically for *Future Scouting*. Worksheets are ideation sheets that generate the artefacts. The prototype is something sharable you create from the artefacts (or you may want to just share the artefacts as your prototype—more on this in Chapter 10).

Download the resources from futurescouting.com.au.

Print the resources out on A3 or load them into an interactive online tool like *miro.com*.

Alrighty then, it's time to prepare yourself for time-travel—get out your **Future Scouting Dashboard** worksheet (Figure 5.0) and meet me on the bridge.

5

Prepare to time travel

Like any emerging design practice that challenges the status quo, speculative design gets us designers as excited as ewoks bringing down a Death Star. Combine this 'new hope' effect with the thrill of thieving artefacts from the future like we're Indiana Jones in a jacked-up DeLorean, the hype can make us forget to keep our wits and values.

In our humble day jobs, we designers do our best to champion the user and guard against dark patterns. But the consumer-driven machine we design for is hungry and moves fast—reactionary timeframes, budget restraints, conflicting stakeholder needs, and tech limitations, often Frankenstein something that could have been beautiful (you're beautiful, too, Franky, in your own stitched-up way).

Use the **Future Scouting Dashboard** to keep your time travelling on track.

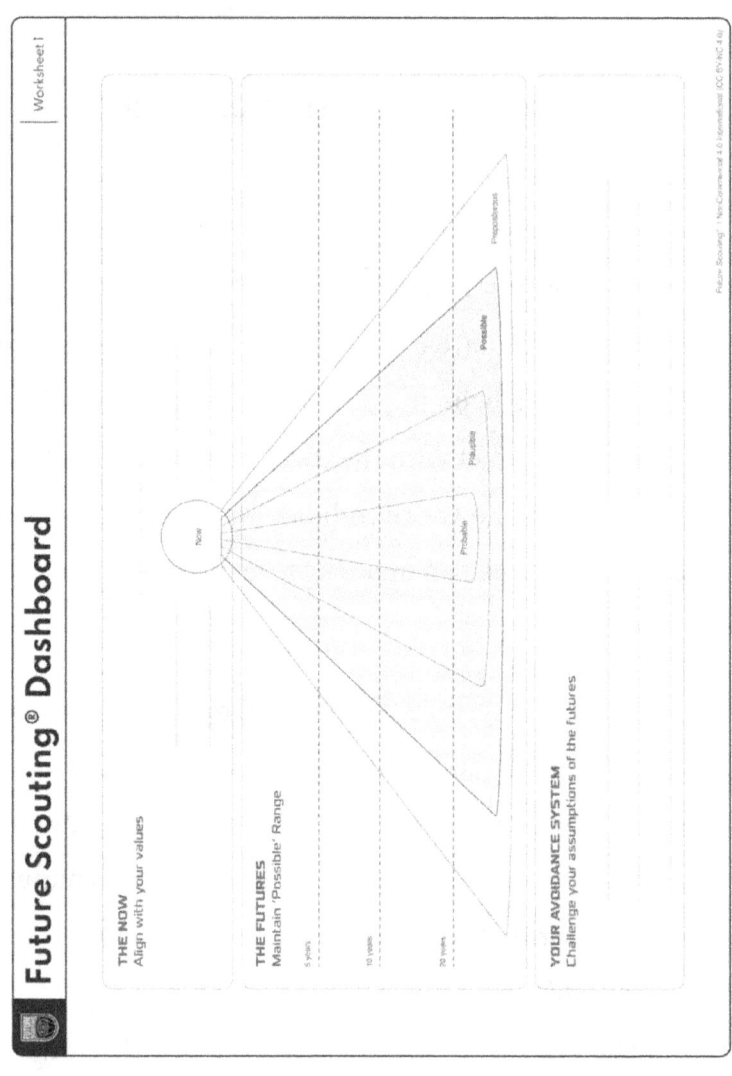

Figure 5.0 - Worksheet: Future Scouting Dashboard

Determine your core values

As mentioned in Chapter 2, speculative design took form in the early 1990s as designers questioned their role in consumerism's impact on the planet. *Future Scouting* aims to champion this foundation and empower designers by embedding core values within the process.

While defining this aspect now might limit the scope of possibility, I'll leave that vastness for other methods. As slaves to the giant profits-focused consumer machine, designers rarely get the opportunity to design according to what is intrinsically important to us—*Future Scouting* gives this back!

Make time to identify your core values to centre yourself and to know what values to champion in the future.

Review the Core Values list at *futurescouting.com.au* to identify yours. Five values are recommended as not too few for us complex beings, yet not too many to be confusing [3]. Write your five values on the lines provided in the NOW section on your **Future Scouting Dashboard**.

If you need help, or just prefer to go deeper, I recommend doing an online exercise such as Scott Jeffrey's "7 Steps to Discover Your Personal Core Values" at *scottjeffrey.com/personal-core-values* or download the flashcards and worksheet from *tomillama.com/personal-core-values.*

When prompted during the *Future Scouting* process, check your **Future Scouting Dashboard** to choose a Key Value. You can use a different Key Value for each *Future Scouting* project.

Know your assumptions

If you web-searched 'the future' six or more years ago, you would have seen a page full of city images with gleaming towers and flying cars. Today, the results are similar, but with the added dimension of 'greened' cities. The predominate available images of the future remain, however, as gleaming, greened, high-tech cities. Stuart Candy and Kelly Kornet drew attention to this homogenisation of available future imagery in their *Ethnographic Experiential Futures* presentation at the 2017 Belgium Design Develop Transform conference.

In response to the lack of prevalent, varying inspiration, Candy and Kornet developed a "design-driven, hybrid approach to foresight aimed at increasing the accessibility, variety and depth of available images of the future" to broaden humanity's perspective of what futures are possible [1].

To help empower your own ability to see beyond the probable, make yourself aware of your own assumptions and bias. Take a moment to clear your mind, and then think about what future you expect to be most probable for the six aspects below. Consider your positive, negative, and neutral perspectives. Record these in the AVOIDANCE SYSTEM on your **Future Scouting Dashboard.**

How do you imagine the future for:

- Social issues
- Technology
- Environment
- Economy
- Politics
- Arts and Entertainment

During each *Future Scouting* ideation session, check your dashboard to see if you've defaulted to any of your assumed versions of the future so that you can go back and ideate harder to push yourself past them. You don't have to change what you first ideate if it feels right—this is more of a sense check and an opportunity to delve deeper into your own creativity and subconscious.

Set your range

Using the futures range on your **Future Scouting Dashboard** as your guide, check your invention remains within the possible future, keeping it realistic yet with room for an unexpected level of change.

Also, try to keep the invention within the near future, no further than the next 20 years or so—this keeps your prototype relatable to your audience. The further out in time we try to imagine our own lives, the less our brain recognises ourselves, and the less it cares [2]. Mark your future time frame in the FUTURES section on your Dashboard.

Time to travel!

You should feel the book humming now, powering up for a warp.

It's about to get wild and speculative, so put your seat belt on because time-travelling can get a little loose.

Remember to check your dashboard as you go to stay aligned and in range.

Ready?

Hold the book tight and... *Warp*!

6
Catch a signal

You're in flight and hurling forward in time!

Actually, the way scientists understand time is that we don't actually move—we cause time and space to move around us. So that gleaming stream of light spears outside your window right now is actually time hurling itself around you.

In any case, you need to get a direction by picking up a 'signal'.

Signals are signs of emerging phenomenon with the potential for causing change. They are flickers in information sources suggesting potential for innovation and/or disruption, whether it's an alternate or extreme amplification of current trajectories. Signals lurk like potential viruses in the fringes of media and discussion, waiting to spark the interest of others who might carry the idea forward into a trend.

Signals can be found in many forms—news articles, social media posts, comments, journals, polls, data monitoring, environmental change, animal migratory patterns, birth rates, research, emerging technologies, behavioural use of technologies, movements, ideologies, attitudes, consumer trends, terminology, and language trends, innovative services, goods, materials, and so on! Signals can be a 'thing' (technology, service, or product) or a pattern shift (new ideology, change in behaviours or attitudes, etc.)

Futurists, governments and noncommercial entities use an ongoing horizon scanning of news, social media and other sources to analyse drivers and trends for change that might establish a signal.

Trends are patterns of emerging signals appearing over enough time to suggest longevity and potential influence on the life of other signals and trends. They're more embedded than a fad, but their nascent nature makes them volatile—full of newly grouped energy with the potential to disrupt, yet young enough to be diffused by other trends. They're slippery buggers, like a pit of trash compactor eels feeding on each other—hard to place a bet on which one will get the fattest.

Drivers are the large forces steering the world—economies, governments, demographics—and are slower and harder to change [1].

With *Future Scouting*, you'll consider trends and drivers as a sense-check on your signal's potential to ensure your invention maintains enough realism to stay within the range of 'possible'.

Time to get your antenna up and start scanning. Pull out your **Signal Tracing** worksheet (Figure 6.0).

A summary of this exercise

- Using the tips below, scan for signals of emerging change and place them on the Futures Cone on your Dashboard to gauge their possibility (example below).
- When you find a signal you're interested in, use Worksheet 2—Signal Tracing to trace/capture detail to understand the signal better, gather information about how often is it appearing, etc, and gauge if it has the potential to grow and cause change. You may

need to trace a few signals to find one that has enough promise
- You might already have a good idea for a signal—feel free to start tracing that.

To summarise what you are looking for:

- A new 'thing' (technology, service, or product) or a pattern shift (new ideology, change in behaviours or attitudes, etc.) or a change in the status quo that, if continues, could result in significant changes in social, technological, environmental, economic, and/or political ways
- A signal that appears in a few instances, maybe in a news article, on a blog, in a comment on social media
- A signal that has some potential for provocation by challenging status quos
- Keep in mind, Future Scouting is about designing a technology-based product/service. If you find a signal that is more like a pattern shift (behaviours,

movements, ideologies, attitudes, etc), you need to focus on the technology involved and a designing a product that fosters or hinders it (or find a more suitable technology-based signal)

Figure 6.0 - Worksheet: Signal Tracer

1. Scan for signals

Scan for signals of emerging change following the tips below and place them on the Futures Cone on your Dashboard to gauge their possibility.

Examples of signals: A mixed reality contact lens, self-driving vehicles becoming the norm, smart clothing, the sharing economy, potential future jobs (e.g., robot lead, food engineer, rewilder, etc.)

Picking up signals is easier for some than for others.

For those of you with an active interest in emerging tech—who already read tech news and follow tech blogs out of your own interest—you're already scanning for signals. Regardless of who you are though, the methods below will show you how to get your antennae on and collecting signals from the future.

Keep in mind as you trace signals, *Future Scouting* is about designing a thing (technology, service, or product). If your signal is more like a pattern shift (behaviours, movements, ideologies, attitudes, etc), you need to focus on a technology that fosters (or hinders) the trend.

If you struggle to pick up a clear signal, use these prompts:

- When do you see the trend happening? Look at futuretimeline.net for predicted technology
- Use the technology and interaction cards at futurescouting.com.au

Desktop research - mine the collective mind

You can research as lean and rapid as you like, but it's important to go deep when doing desktop research, as a shallow scan will usually reveal only a common view of the future.

Look for what surprises you and sparks your curiosity. Keep an eye out for signals that grab your personal interest, and ideas or comments that make you pause and wonder 'What if?' You'll be spending time researching this signal, so the process will be more fun and interesting if you're researching something that naturally engages you.

The Tracing Tips below are some of my own combined with advice from professional foresight practitioners

[2][3] and have been combined with ideation models such as STEEP and the *Seven Foundations of Worldbuilding* [4]. Follow these and you're sure to spot a flicker of a potential underrated future. Track your exploration by recording links, notes and screenshots on your **Signal Tracing** worksheet (but don't feel restricted by the layout—map out your own layout if you need to).

If you're like me, and you record a lot, highlight notes that really stand out as you go. Check article dates to avoid outdated signals and keep an eye on the quality of sources.

Tracing Tips

- Start with either a broad search term, like 'emerging technology', or 'the future of...' any industry, country, idea, value, cause, etc., that you're interested in
- Remember, if you spot a signal that is an emerging change in the behaviors and/or attitudes of people, animals, the environment, or the economy, look for the technology involved

- Follow trails, look into comments
- Cast a wide net of curiosity and mix your sources and perspective, from personal blogs to professional news sources, just look for well-argued opinions and check any sources:
  - Research companies (Gartner, CB Insights)
  - Tech/innovation/design/science news sites and blogs (The Verge, Medium)
  - Social media
  - Tech and innovation sites (Mashable)
  - Tech and innovation sections in news sites
  - Chat/discussion channels (Reddit and Quora)
  - Innovative design groups
  - Innovative designers, leaders, and thinkers on Linked-in
  - Academic journals
  - Industry specific information streams
  - Artists and writers producing speculative work
  - Conferences
  - Niche bloggers

- Explore beyond the first page of Google results
- Read random articles, explore the ridiculous
- Drill down into user comments and listen to demographics, cultures, and perspectives outside of your own
- Download trend reports from:
    - Future-focused websites (iftf.org)
    - Research and insight companies (CBinsights)
    - Advertising and design agencies offering free information reports and packs
- Watch and listen to conference videos and podcasts (CES and DLD20)
- Scan investment reports for emerging market ideas
- Enjoy exploring visual data infographics like those at informationisbeautiful.net
- Kick back and read the synopses of the most recent sci-fi films and TV series

People watching

Get away from the black mirrors of your laptop, TV, and mobile, and go out into the world. Take note of what's happening in your immediate community while you're out shopping, on public transport, in cafes, etc.

Visit areas you don't normally venture to. Visit culturally diverse areas and tourist areas. Visit a few different areas or just one.

Sit and observe people at work and leisure, on weekdays and on weekends. Are there any behaviour trends you hadn't noticed before? Look at small things, and let your imagination run with them.

For example, are more people looking down at their phone while walking without needing to ever look up? Are we developing our peripheral vision? How would that new norm affect social interaction?

Use your mobile to capture any thoughts, questions, and ideas that pop up, or write them directly into the first row of your **Signal Tracer** worksheet. Continue with desktop research to assess their potential.

Survey your communities

Another method to capture thoughts and ideas from others is to use brief, anonymous surveys created through tools like Survey Monkey or Google Forms. These are useful if you have a particular signal in mind you want to explore.

Fast and furious

If you struggle to find a satisfying signal and just want to get started, you can use the *Future Mission Tool* at *futurescouting.com.au*.

2. Trace a signal of interest

Signal Tracing worksheet as a guide to know what to ask to better understand the signal's potential and draw out its detail.

When you spot a signal, keep researching and use the Signal Tracing worksheet to collect links, notes, and screenshots. You don't have to completely fill out the worksheet, it is just a guide to better understand the signal's potential.

Keep in mind as you trace signals, *Future Scouting* is about designing a thing (technology, service, or product). If your signal is more like a pattern shift (behaviours, movements, ideologies, attitudes, etc), you need to focus on a technology that fosters (or hinders) the trend.

Collect links, notes, and screenshots. You don't have to completely fill out the worksheet, just use these as a guide to gather information about the status of the signal and to check it has potential to grow and cause change:

- Practical drivers—look for drivers that spur on the need for its adoption:
    - Technologies and innovations
    - Environmental changes
    - Economic shifts in power
    - Political climates
- Cultural drivers—scan the segments below to find recent variations of your signal as a sign it has potential to become a trend:
    - People (from teenagers to middle age to the elderly)—are people talking about it? does a related post get a lot of

engagement (likes, dislikes, comments, etc.)
- o Cultures (from race, sexuality, gender, etc.)—are you seeing variations, mentions, or adaptations across cultures?
- o Entertainment—has the concept been represented yet in any art, media, pop culture, or music, etc.?

- Assess a signal's adoptability
  - o What might foster adoption of the signal?
  - o What might hinder adoption? Keep an eye out for anything that might make this impossible. But also keep in mind your future range on your Future Scouting Dashboard—if adoption requires technology yet to be developed, check if this development looks likely in the near future, as it may not be an issue. Remember this is speculation, so don't dwell on accuracy, use your best judgement, be daring, and keep it fun

- Let your search for a signal be an organic process—you might notice a driver first and then drill into that for a signal. Don't overcomplicate it. Find something that catches your imagination and run with it
- Capture any random questions and ideas that spark during your search—this is the beginning of your future-world building
- You could trace several signals and then choose the most potent or interesting

Summarise with story

When you decide on a signal with potential, write a short summary of its story. Use the basic start-middle-end structure to assist your writing:

- How did the signal originate?
- What is its current state?
- What are its potentials?

Signal Tracing Example

Below is how my thought process flowed as I traced the signal of augmented reality filtering our daily vision as much as we filter what news we read and what we share on social media.

From a personal interest in augmented reality, I stumbled across Snapchat Spectacle AR 'filters'. While these filters and effects are applied via an app *after* the spectacles captured the video, the concept of in-vision filters altering a daily view—and how this might affect the wearer—sparked my curiosity.

Exploring the concept of AR filtering our daily view beyond gaming and or task-based activities my line of tracing went something like this:

- I scanned existing AR glasses available for purchase to learn the current limited state of consumer AR, and then I scouted tech forums, sites, and conference videos to check how far the technology was being pushed. Smart lenses were in development, but far away from anything super immersive, as there are major technical problems to overcome—limited field of vision (FOV), realistic resolution, and the power needed for constant spatial tracking to display full mixed reality

- Hand-tracking was evolving, but by this stage I was imagining a wearer controlling the AR with their eyes from a lens, so I researched eye-controlled interfaces of which there was little for consumers
- Looking for potential hindrances and drivers, I discovered great enterprise investment was driving the development of AR visors, and advances in eye/vision control technology was driven by the need to help the less abled. Consumer demand for more realistic and immersive AR and MR was unmissable.
- Exploring adoptability, I looked at Google's early Glass innovation and how privacy and fashion had been an issue, but also how the fashion aspect had been greatly improved by AR glasses I saw on Amazon

Looking at articles about the mental health impact of the constant filtering our own photos, and the social impact of self-curated news and information, I could see there was potential for dark design trends being used, and damaging behaviour trends to arise. But the potential—a fully filtered augmented vision—clearly held undeniable excitement.

This combination of hype and danger was something that could be leveraged in a future invention design process to utilise irony and humour, and to provoke emotion and challenge thinking—in summary, a strong signal.

Next

The next step is to start evolving your signal into an artefact. You'll research, ideate, and sketch a concept into life that will enable your Key Value.

You're only halfway through your journey through time and space.

When you're ready, top up the carbon-freezing liquid on your ship—it's time to hunt down that future invention.

7
Hunt an invention

Your tracking device is homing in.

Open up your **Research Map** worksheet (Figure 7.0) and complete the mission statement at the top by writing in your future tech signal in the place provided.

Now align your mission with a Key Value—choose one of your Key Values from your **Future Scouting Dashboard** to champion and write this in the place provided on your **Research Map**.

Ok, let's get hunting.

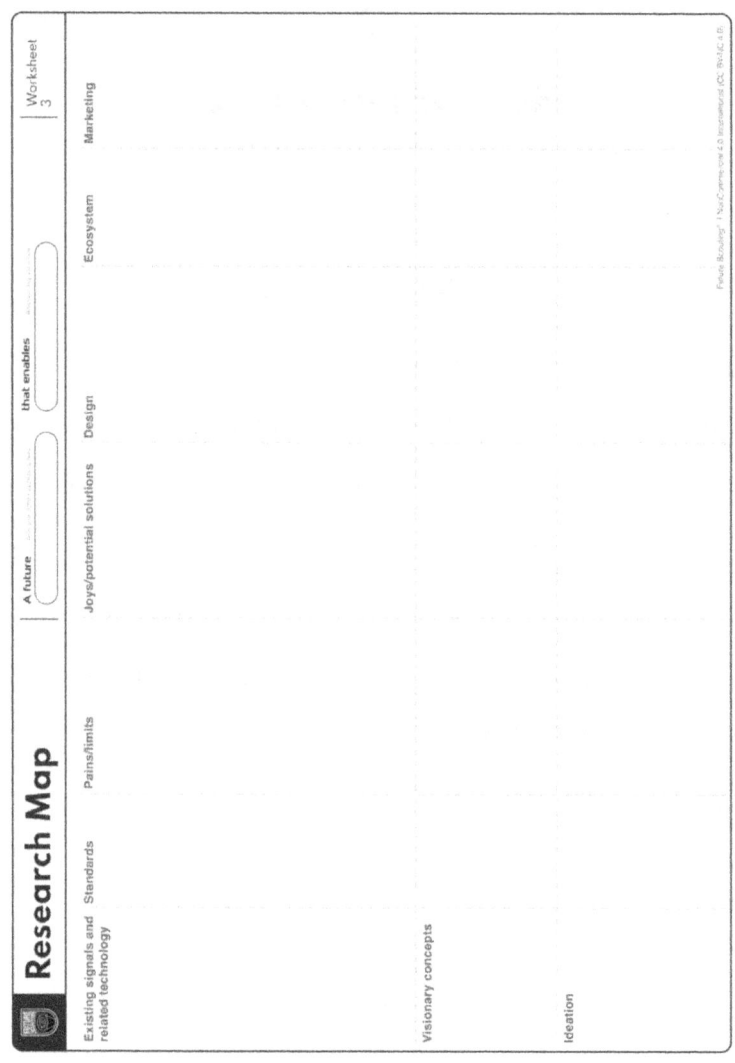

Figure 7.0 - Worksheet: Research Map

1. Ground your invention in realism

As mentioned in Chapter 4, *Future Scouting* grounds speculative designs in realism to help audiences suspend their belief and to help you, the *Future Scout*, keep within the 'possible' range. Research the variations of your signal—from existing and emerging to future concepts—and capture the insight on the **Research Map**.

Start by web searching for an existing product—or different products that use similar technology to your signal—and make a note of the following:

- **Standards**—compile a list of features that might be standard/necessary features (like the default controls and apps on a smart phone—settings, volume, brightness, etc.). *Where to find: product feature lists, FAQs, tech specs, customer reviews.*
- **Pains/limits**—What are the technical, ethical, or other limitations that manufacturers and users hope are overcome. You might come across some products that have attempted some solutions - note these, too, in the joy column, to iterate on later.

Where to find: FAQs, tech specs, customer reviews.

- **Joys**—features unique to certain products, their benefits, the needs they serve. *Where to find: product feature lists, FAQs, customer reviews.*
- **Design**—note any visual design treatments and/or product-specific design principles that catch your attention
- **Ecosystem**—Does the invention require connectivity or rely on other devices? How are customers supported (FAQs, tech support)? Does the invention have a fan community—do they modify it? (comment sections in sites like Reddit are a great source of this); do fans congregate online or in-person?; what terms do they use?
- **Marketing**—slogans, promotional terminology, etc. provide inspiration for your design later on
- Anything else that inspires storytelling

Repeat this for as many variations of the signal and/or related technology as you want, but be sure to explore:

- Emerging/bleeding-edge technology and innovations
- Visionary concepts found in entertainment media (TV, film, novels, comics, games, art, etc.)
- Mock products in comedy that often highlight the product challenges and irony of its use/purpose/impact
- Cultural variations

Research Map Example

Below is how my thought process flowed as I researched the advanced augmented vision signal from the previous chapter.

- **More signal variations**—I included and expanded on what I discovered during the signal tracing
- **Details of related existing tech**—AR eye wear, MR versions, smart lens technology
- **Ecosystems**—connectivity and companion devices, such as power packs and connected phones and watches. I thought about what support lens wearers might need, such as eye

care and 'vision' bars like Apple's Genius bars but with optometrists

- **Conceptual**—Avenger's Iron Man HUD, Magic Leap's concept art, and Artist Keiichi Matsuda's Hyper-lapse future AR concept video—youtube.com/watch?v=YJg02ivYzSs—were great starting points for device features and interface design ideas
- **Related technology**—Eye-controlled interfaces for those who cannot move or speak. Thinking in terms of transitioning this tech to consumer products, there appeared potential for new interaction methods to be designed

As with signal tracing, let your search be organic, follow leads and tunnels of interest until you collate a healthy mix of information.

If your own ideas pop up as you research, capture them in the ideation column.

2. Evolve your invention

Time to get creative and ideate new features.

Now you'll invent new potential features for your future invention. You'll decide on which ones to use later, so for now the key thing to remember is that this is meant to be playful. Just have fun and don't judge your ideas before exploring them. There is no right or wrong—controversial and ambiguous ideas can be provocative later. As a guide while ideating, remember to:

1. Defer judgment
2. Strive for quantity
3. Seek wild and unusual ideas
4. Build on other ideas

Following the guides above, and using the methods and tools below, ideate new and advanced features and capture them in the ideation column on your **Research Map** (and print more copies if you need more room).

2.1. Warm up your creative brain

To enter a playful mode, you'll need to turn on your right brain (the creative one). Try this simple exercise from gamestorming.com. You'll need a pen or pencil, a few sheets of blank paper, and a timer.

Squiggle Birds [1] activates the pattern recognition capabilities of your brain and will also get you comfortable drawing for the sketching session ahead.

- On a piece of paper, start by making 10 random squiggles
- Start a timer for a few minutes
- For each squiggle, add a beak, eyes, and feet to create a little bird. You can place the bird elements anywhere on the squiggles—the head may be at the top, on the side, or at the bottom as if the bird is bending down
- Keep going until time's up

When you're done, your brain will be in play mode and you'll be primed for inventing!

2.2. Ideate new features

Like signal tracing, ideation comes more naturally for others. Use the prompts below to jump start your imagination.

Ideate joys from new technology

Using the Emerging and Visionary concepts on your **Research Map** as inspiration, imagine new variations—go as wild as you like!

Shuffle and flip the Technology and Interaction Cards at *futurescouting.com.au* .

Use these prompts individually or combine them. If one doesn't feel relatable to your signal, shuffle and flip again. Do as many or as few as you like.

Pains/limits

Ideate how the innovative and advanced ideas above might solve or improve on the pain points and limitations.

- How do these solutions become new joys?
- What new features might be needed when these limitations are reached
- Use the visionary concepts you recorded as starting points to ideate new variations

Consider inclusivity

There are at least 64 genders and 46 types of sexuality [2], over 3800 different cultures and more than 6900 languages [3], and at least 7 types of disabilities [4].

The users of your product could experience any combination of these—and considering we all age and lose abilities, "we are all only temporarily abled" [5].

- Using the Inclusivity Cards at futurescouting.com.au/, ideate ways your invention could include people who are often marginalised by mass design
- For example, if a feature requires sight or hearing to experience it, how can you adapt the design to be operated by visually impaired people, remembering there is a spectrum of visual ability? Or, if it's a wearable, how can it be worn by all body types?

Consider environmental impact

The Environment Cards at *futurescouting.com.au/* consist of six sections that follow the true lifecycle of a product, from where and how its materials are sourced, through the manufacturing, supply, and use stages, through to what happens at the completion of its usable life and the actual breakdown of its parts back into the natural world.

- Use the Environment Cards to brainstorm how the production of your invention might impact the world
- Ideate solutions. Feel free to do some more desktop research, but don't get bogged down in detail
- Look at futuretimeline.net for ideas for your invention's time frame
- Check your invention's drivers to see what other affects these might have on the world, and ideate how this may create new uses for your invention

2.3. Ideate new design principles

Do any of your innovations above require new thinking in regard to design and/or interaction?

2.4. Explore an eco-system

To aid the future world-building that you'll do later, ideate for any potential eco-system of products and services related to your invention:

- **Materials**—How can the materials be improved? What new materials might be innovated? Look at futuretimeline.net for ideas for your future time frame
- **Connectivity**—Does your device need connection with the internet, networks, or other devices? What companion devices might your invention need?
- **User Support**—How are your invention's users supported—online, in-store, by phone, holographic virtual assistant?
- **Community**—Does your invention have a fan community? Do they modify it (Reddit is a great source for this)? Do they congregate online or in-person? What terms do they use?

2.5. Marketing

Using your marketing research, brainstorm new slogans andtaglines based on your innovations above. These don't need to be polished—you'll refine and choose the best later. But these are good to do now because articulating ideas in snippets can generate more ideas to innovate on.

3. Converge best ideas into one invention

Your tracking device is going off—your invention is so close!

Review your **Research Map** and highlight:

- Ideas that relate to your Key Value. Championing the Key Value may generate design that marginalises other values, and this may seem counter-intuitive, but this is what speculative design is about—generating discussion about the directions we aim for and how they might lead us somewhere else
- Controversial or dystopian features—irony and provocation add dimension and engagement to your invention

- Make sure to check your dashboard to stay in range and avoid assumptions

If you highlight more than one version of a similar feature, see if you can merge them, or choose the one that best works with the other features.

4. So what does it look like?

It's time to visualise your invention with some simple sketching. You'll need a pen or pencil, a few sheets of blank paper, and a timer.

If sketching doesn't feel natural for you, or you just want to warm up more, try another *Squiggle Birds* session first.

- To start, read through the features and ideas you've highlighted on your **Research Map**
- When you're ready, set your timer for 10 minutes, and start sketching as many variations of your invention as you can. If you exhaust your creativity before the 10 minutes, that's ok, but try pushing yourself before giving up

- When you're done, highlight which parts of your various sketches you think are the best

5. Blueprint the final design

There it is—the invention is right in front of you!

Using the **Invention Blueprint** artefact, compress your invention into a snapshot:

- Sketch a final version of your invention that incorporates the highlighted parts from your previous sketches that best work together
- Sketch or note any key elements of its ecosystem (companion devices, batteries, support systems, etc.)
- Summarise your invention's purpose by using the Product Vision Statement prompts on the **Blueprint**
- List your final selection of standard and key features, and ideate the user needs from these

Congratulations—you just stole a product from the future!

Time to see how the existence of your invention might impact a future world.

Don't forget to check the progress tracker on your **Dashboard** if you start to feel lost during the process—time travel can make things wobbly!

8
Release your invention

Now you hold your invention in your hot little hands, you'll extrude a future world from within.

You'll start by exploring the potential impact of your invention's release into the world using the **Scenario Futures Wheel** worksheet (Figure 8.0).

About the Futures Wheel

Invented by futurist Jerome C. Glenn in 1971 to explore the consequences of change, the original Futures Wheel has evolved into a decision-making and idea-exploring tool [1]. The Wheel gave structure to brainstorming (recalling what you know about a topic) and mind-mapping (working out relationships between those points).

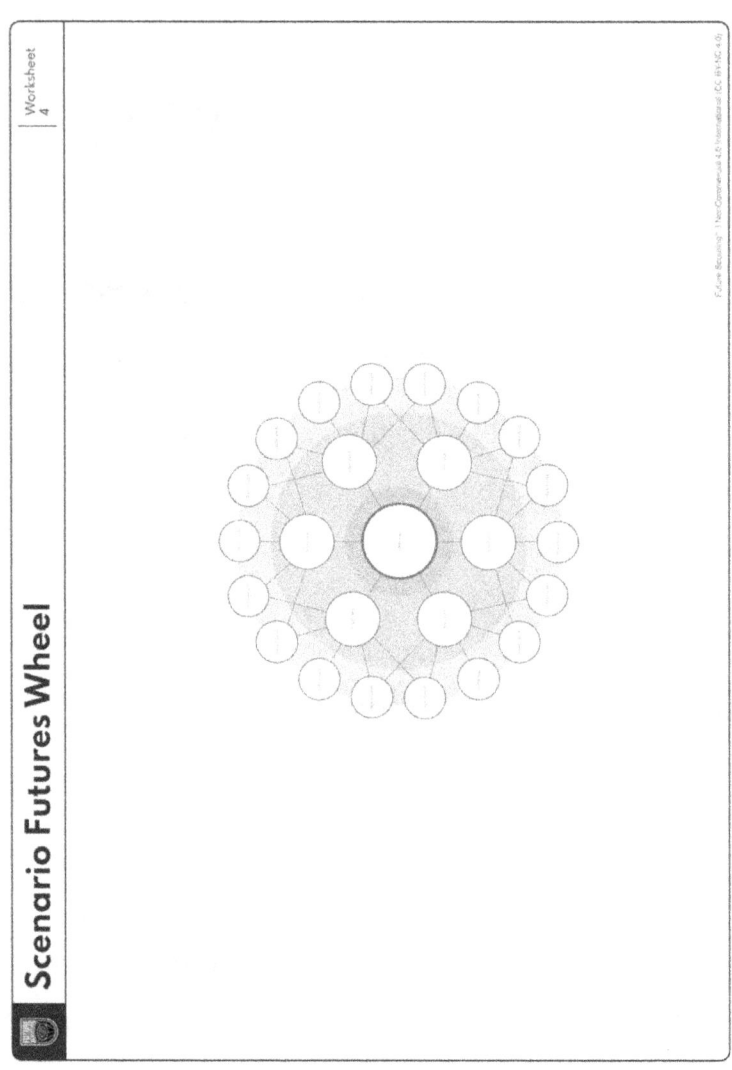

Figure 8.0 - Worksheet: Scenario Futures Wheel

Concerned his original wheel didn't force users to consider consequences beyond the economic impacts, Glenn created a Version 2 that segmented the wheel into predetermined sections including cultural, psychological, welfare, technological, educational, political, environmental, and economic. He then created a Version 3 to consider time.

Using Glenn's version 3 as inspiration, *Future Scouting* uses the original version in conjunction with the Inclusivity, Environment, and Mis-use Cards at futurescouting.com.au/. Using the original Futures Wheel allows for more free-form ideation on the wheel, while the cards provide prompts for consideration factors provided by Glenn's Version 3.

Launch the invention into the world

Step 1

On the **Scenario Futures Wheel, w**rite your invention concept in the centre circle labelled 'Invention'.

Step 2

Imagine this product is released on a large scale and brainstorm possible direct consequences. These can be positive, negative, or neutral. Write them in the first ring around your invention. You can also explore questions and problems to help generate various scenarios. Use the Inclusivity, Environment, and Mis-use Cards as prompts:

- Inclusivity cards—how might the scenario exclude some groups and what impact does it have on them?
- Product lifecycle—how might the scenario impact the environment or non-users?
- Mis-user Cards—how might the invention be abused, hacked, or modified for a different use?

Step 3

Identify indirect consequences generated by the direct ones. Use the connecting lines to help you think of what results might occur from the combined effects of two other results. These lines are just a guide, feel free to ignore, add, or remove as needed.

Tips

- Be specific with your ideas. For example, if one of your results is "Increase in wearable use", be more specific about what technology, and perhaps even about the frequency, such as 'Smart watches worn 24/7'
- Feel free to move direct results around to explore different and more exciting indirect results
- Explore the weird, controversial, and dystopian as much as the idealistic concepts
- If you get stuck for ideas, use the following prompts to spark your imagination:
    - Look at user needs on your blueprint
    - Flip a Future Scouting card
- Desktop research the more technical ideas - be as in-depth or as rapid as you like

Step 4

Extend out into more indirect results as many times as desired—go into third or higher levels of consequences.

Step 5

When you're ready, identify the ideas most related to your Key Value, and aim for ones that might cause people to stop and think. Don't forget, even seemingly mundane scenarios can be provocative, particularly if they juxtapose a familiar experience from a future-twisted perspective. Highlight your favourite—this is your Key Scenario.

Tip

You could also brainstorm a wheel focusing on one interesting feature that you want to explore. You can do one or as many wheels as you like but be warned—if you're excited by future concepts, these wheels can become addictive!

Meet the hero

Now you have a Key Scenario, it's time to meet that scenario's hero and discover the holographic universe hiding in your invention.

9
Hero and future world

Welcome to your future world!

With your invention's Key Scenario revealed, extract a persona for your scenario's hero(s) and then zoom out to see their world.

The Hero

In 1962, Hanna-Barbera Productions introduced America to The Jetsons, a utopian future cartoon about a family living in a sky home, getting about in flying cars, living with robot assistants, and enjoying the extra leisure time provided by three-day workweeks and a menagerie of automated conveniences.

While the hero of this cartoon was George Jetson, the father and head of the family, the other characters and their micro-stories gave us various glimpses into this possible future—Rosy the robot maid's 'relationship' with another robot; Astro the real dog with his rudimentary grasp of English; and George's free-thinking work computer R.U.D.I being a member of the Society Preventing Cruelty to Humans.

While the key user of your invention might be considered the 'hero' in a sci-fi film about your invention, the above characters are the types who might be the hero of your Key Scenario—those non-users and non-humans who are impacted by your invention in an indirect yet interesting and provocative way.

Look at the Key Scenario you highlighted on your **Scenario Futures Wheel**—who or what is at the centre of this? They might be an AI, some aspect or ecosystem of the environment, a technological network, or a homeless person. This is your Hero.

Identify your Hero's persona type from the list below, and create their persona using the **Future Persona** artefact (Figure 9.0).

There may be several characters in this scenario, so feel free to persona them all:

- **User**—the key user of your invention

- **Non-user**—a human who doesn't use your invention but is impacted by its existence at some time during its lifecycle (e.g., a worker in the manufacturing plant; a child forced into labour to dig the invention's raw materials from the ground; a farmer whose land is impacted by the inappropriate disposal of your invention)
- **Non-human**—an animal, insect, plant, ecosystem, ocean, migration pattern, etc.
- **Business**— the organisation that might own the invention, or a business impacted by your invention's existence at some time during its lifecycle (eg., a company that manufactures parts for your invention, a company put out of business by your product)
- **Mis-user**—someone who uses your invention in an unintended way, such as a hacker, terrorist; but this can also be someone who modifies it to use for a different but benevolent cause

Personas

Refer to your **Invention Blueprint** as you answer the following questions to complete your **Future Persona(s)**.

When answering these questions, try to answer for the persona, whether human or non-human, from their perspective using the first-person voice.

Keep in mind these personas relate to how they are *before* they engage with—or impacted by—your invention.

User

- Imagine this persona in their safe space (home, room, shelter, cave, vehicle, etc.)—describe it and the persona's relationship with it
- Describe what they look like, their gender, age, what they're wearing
- Give this person a name
- How is the key value important to this persona?

- There are two photos on the wall that shows this persona's life/past—one at a social gathering, and on holiday. Describe these scenes —the environment, things and people—and the emotion they generate in this persona

Non-user

- Imagine this persona in their safe space (home, room, shelter, cave, vehicle, etc.)—describe it and the persona's relationship with it
- Describe what they look like, their gender, age, what they're wearing
- Give this person a name
- How is the key value important to this persona?
- They keep two photos/images with them—one of the past and one of a wish. Describe these scenes —the environment, things and people—and the emotion they generate in this persona

Non-human

A non-human can be an island, a plant, air quality, an AI, or other. When answering these questions, try to answer for your non-human from their perspective using a first-person voice:

- Describe what they are, what ecosystem or network they are part of and what they do
- Give this persona a name
- How is the key value important to this persona?
- If they are important to the invention, how?
- If they are important to environment, how?
- If they are important to social or other systems, how?

Business

The organisation that might own the invention, or a business impacted by your invention's existence at some time during its lifecycle (e.g., a company that manufactures parts for your invention, a company put out of business by your product)

- Describe what they are, what they do, and their relationship to the invention
- Give this persona a name
- How is the key value important to this persona?
- What are their hopes and fears for the invention?

Mis-user

- Imagine this persona in their safe space (home, room, shelter, cave, vehicle, etc.)—describe it and the persona's relationship with it
- Describe what they look like, their gender, age, what they're wearing
- Give this person a name
- How is the key value important to this persona?

Figure 9.0 - Artefact: Future Persona

- There are three photos on the wall that shows your user in the world outside—one at work, one at a social gathering, and on holiday. Describe these scenes—the environment, things, and people—and the emotion they generate in the user
- They're waiting to get their hands on your invention, how are they looking at it—through a window, on a screen, outside, etc?
- How is it making them feel?
- What are their intentions with the invention?

Key Story

Now you know your Hero, start zooming out to see them using the invention and their immediate world.

To start, use the **Story Futures Wheel** worksheet (Figure 9.1) to ideate a Key Story that will form the basis of your future story for sharing back to an audience.

On the **Story Futures Wheel,** write in the centre how your hero initially engages with—or is impacted by—your invention. Give context but keep it simple. For example—my Hero activates his Lenz as soon as he's outside the store.

Firstly, for the top half of this wheel, brainstorm the different ways your hero interacts with your invention. Then brainstorm the effects of these actions, either on your hero, the invention, others, or the world.

- What are their intentions with the invention?
- How is it making them feel?
- How does the invention enable or hinder the key value for them?
- Does the user have any of the other ecosystem parts? What are they doing with them?
- Does the invention connect to any social/multi-user networks, and how does that influence how they engage with it?
- How does their use impact others?

Continue to think in terms of cause and effect. Follow leads and extrude into snippets of story. Play with this until you have one or more story ideas. Highlight the one(s) most related to your Key Value.

Next, move to the bottom half of the wheel and imagine your user has been engaging with your invention for six months or a year or longer.

Using the highlighted snippets from the top half of the wheel, brainstorm your hero's long-term habits with the invention, both uses for it or behaviours and reactions to it.

Again, follow leads and extrude into story snippets. Continue stories from the ones you brainstormed in the top half—take them to a climax and then resolve (with a positive or negative outcome, it doesn't matter).

When you're happy with one or more story ideas, highlight those related to your Key Value. Group any snippets following the same story by drawing one line around them all.

Take a moment to review your creative brilliance!

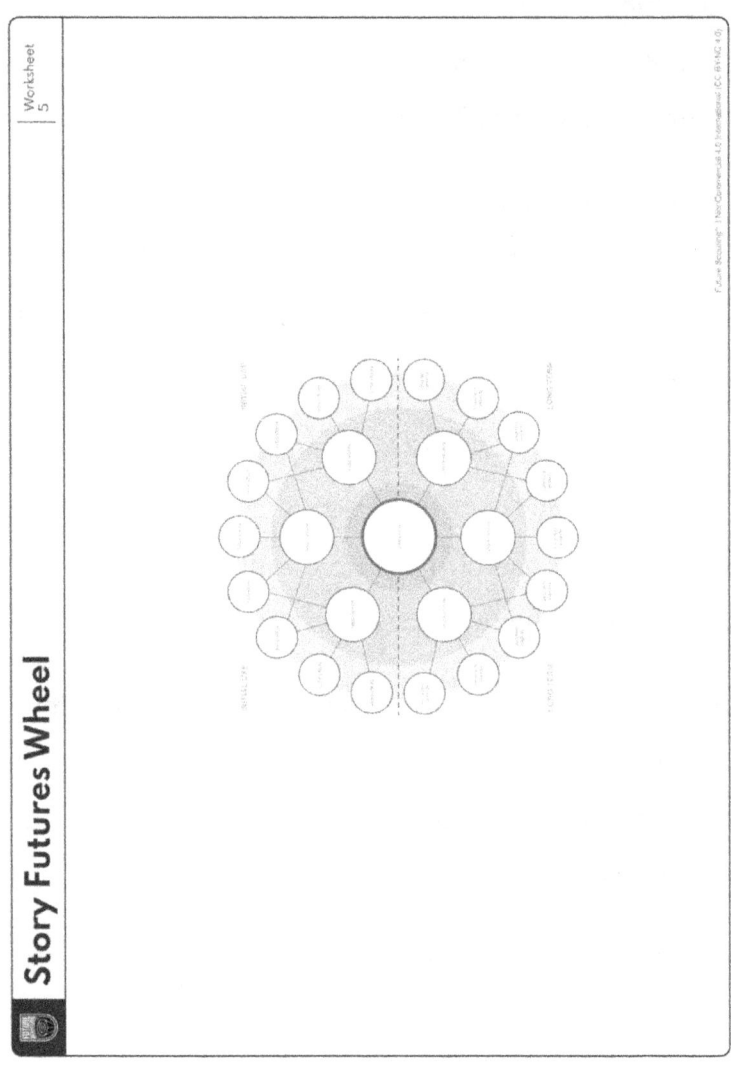

Figure 9.1 - Worksheet: Story Futures Wheel

Which snippet—or group of snippets—moves you the most. It doesn't have to be the most exciting or action-based, it can be quiet and simple, but by its ironic nature stirs the most conflicting thoughts. Highlight this as your Key Story.

That's all you'll do with this worksheet for now, but you'll use it as inspiration for when you create your prototype.

Before you move on to the final artefact exercise, reflect on what you've discovered, and update your **Invention Blueprint** and **Persona(s)** if needed.

The future world setting

To capture extra setting details and mood for your prototype, zoom out further to discover a wider view of your invention's future world.

Generate a snapshot of your invention's future world by using the tips and prompts below to complete the **Location snapshot** artefact (Figure 9.2).

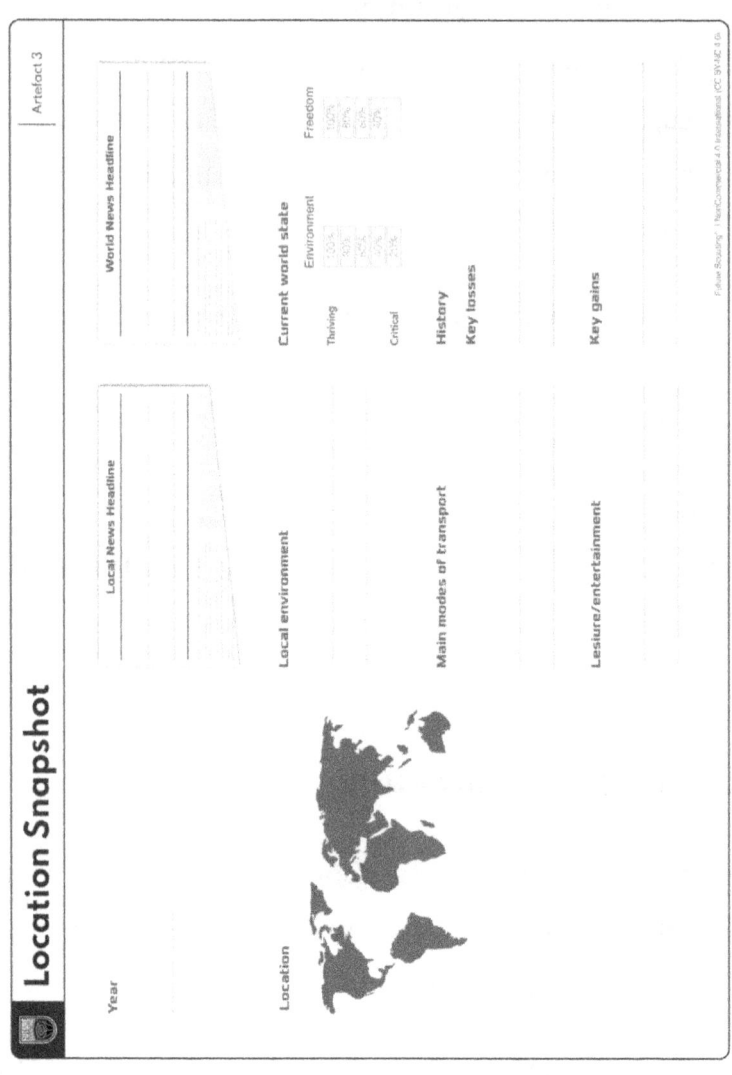

Figure 9.2 - Artefact: Location Snapshot

Tips

- As well as the prompts, draw from ideas on your **Persona(s)** and **Story Futures Wheel**
- Refer to the Drivers on the **Invention Blueprint** to ideate the 'History' events and how they formed the 'Current World State'

World-building prompts

- What year is this? Refer to futuretimeline.net/ to help you ideate what happens or exists at this time
- Where are they? What visual noise is there, like brightness, movement, large signage, augmented information overlaying the world, etc.?
- What's the weather like? How does it affect the area—what is the quality of the environment, is there much green, how is the air?
- What are the main modes of transport?

What's the state of the world?

- Looking at the drivers on your **Blueprint**—what changes did they drive between now and this future?
- Describe any key events that lead to this state
- How did these events affect the environment?
- How did they impact society?
- What/who thrived and how?
- What/who suffer and how?

Time to head back to your Now

You've done it—you've captured an invention and collected snapshots and artefacts from a possible future!

You're light speeding through the wormhole back to your Now, enjoying a chilled glass of blue milk blue and watching the gleaming light stream passed the windows. You've accomplished a lot, seen a lot of wild things, and you deserve this moment of rest to reflect on what you've collected.

- **Invention Blueprint**
- **Persona(s)**
- **Key Scenario and Story**
- **Future Snapshot**

You've done well.

But you're not done yet.

The final step in *Future Scouting* is the most important.

What form you choose to share back your prototype of the future will affect how well the universe of information contained within will inspire people from the present to create the future they really want—or at least avoid the ones they really *don't* want.

Meet you back at the cantina. I'll have a *Greedo's Lament* waiting for you.

10
Prototype and share

This step is a means of refining your concept one final time, bringing all the artefacts together into a design fiction.

Let's start with remembering the purpose of your journey into a possible future—to generate discussion, debate, and awareness, beyond projected or plausible futures, so that designers, companies, and the public, not only live more aware of how their actions contribute to manifesting and hindering the future, but so they also begin to imagine and articulate their preferred futures.

To achieve the above, you could review your artefacts and hold back what should be avoided in the future and what should be nurtured. But while it's fine to promote what you truly believe is right, this could be perceived by your audience as preaching based on your own bias and your own preferred future—especially when their idea of a better future is different to yours—causing your audience to stay closed to your invitation to suspend their belief.

You are a journalist, of sorts, reporting from the future through design, so your story must be as free from your bias as possible. While this journey has been based on your own personal values, how you present back the findings should be impartial, so that the good and bad implications of an attempt to focus on the Key Value can be experienced by all.

Use restraint in the detail you include so the audience's imagination is sparked by filling in the gaps themselves. Show the audience glimpses of the product's potential by choosing the most mundane and provocative aspects and combining them in a confident way.

Option 1 – Package your product

To get you on your way, use the packaging template in **Artefact 4—Packaging** to create a visual of your packaged product. You can then submit this to the forum to be 'stocked' in the ALT-FUTRS store (futurescouting.com.au/alt-futrs).

The ALT-FUTRS store is a design fiction itself, a future online store where consumers of today can browse the latest gadgets and technology from the multiple tomorrows, from holographic holiday escape passes and body suits that harness body energy to mixed reality projectors for the home and headphones that interpret the earth's energies into sound.

Use the template in **Artefact 4—Packaging** with your preferred illustration/mock-up software (e.g. Illustrator, Figma, etc.) or online collaboration software (Miro, Mural, etc.).

- Load the template into Figma, Miro, etc.
- Use text, royalty-free imagery, and icons to visualise the packaging for your future product. See the included example for inspiration and follow these suggestions to help you create the content:
    - The front—Image, name, tagline, any key feature/benefit
    - Side 1—The promotional paragraph, what it is, the problem it solves, benefits
    - Back—list all features and/or benefits or technical specs

- - Side 2—Your design challenge, the technology and Key Value you designed for, add your name and the year your product exists in

- When you've completed your packaging, email a high-resolution image of your Packaging Artefact to futurescouting@outlook.com and it will shortly be stocked in the ALT-FUTRS store.

Option 2 – A case study for your folio

Combine your worksheets and artefacts into a speculative design case study for your folio.

Show the story of your process with the artefacts as your output.

Take the viewer on a journey through time and show prospective employees and collaborators that you experiment with design and think beyond today.

Option 3—Get creative

Explore other ways and mediums to use to package your vision. Think in terms of these factors:

- **Prototype mediums**—the 'how' (exhibition, workshop, themed event, etc.)
- **Playback channels**—digital or physical or a mix
- **Content**—the objects you make to include
- **Interaction mediums**—how to engage audience participation
- **Point of view**—will the prototype 'feel' as if it's shared from the hero's perspective, come from multiple perspectives, or be a silent narrative?
- **Tone and language**—in general, aim for a protopian tone for relatability, as anything too dystopian or too utopian can sound contrived
- **Echoes**—how to perpetuate the playback

Mix and match the elements as much as you like—your playback could be one artefact or several; all digital or a mix of physical and digital; a digital thing in a specific physical space; be at one time, over several days or weeks; or online forever.

Mix and match your approach—one element will affect what interaction options you can choose from, so you might start with a preferred interaction medium and work back from there.

Below are examples of the six factors above to get you started—but this list is by no means fully comprehensive.

Examples of mediums

When choosing the medium, consider what represents your future world, or creates an interesting juxtaposition with what might be expected from your future world.

Type	Example
Exhibition	Gallery
	Online virtual gallery
	Pop-up space
Guerrilla installation/tactical media *Please note, anything that misleads an audience in the virtual or real-word environment, or affects someone else's or a public space, must be treated with appropriate consideration, respect and care.*	Placing your future prototypes amongst real products for sale in a store
	Posters of speculative scenarios, mailing out postcards from the future
	Radio drama
Faux promotion *Treat with the same appropriate consideration, respect and care as with Guerrilla installations*	Billboard poster
	Company/product website
	Digital advert
	Posters
Themed event	
Workshop	
Game	
Story	Short story, book
	Case study
	Comics/Graphic Novels
	News article/ headlines /breaking announcement
	Video
Performance	Character video
	Documentary
	Mockumentary
Theoretical	Diagram flows
	Theories

Figure 10.0 - Table: Prototype mediums

Playback channels

- Physical thing (e.g., sculpture, physical product prototype, etc.)
- Digital thing (static web page, video, email, digital prototype, etc.)
- Physical experience (exhibition, guerrilla installation, performance, etc.)
- Digital experience (web site, online game, online survey, etc.)

Content

When you know the channel and medium, take time to choose the content for your prototype that will convey the invention, Hero, and Key Story. Review the details in your **Story Future Wheel**, **Persona(s)**, and **Location Snapshot** for content you might need to make, illustrate, write about, or other.

Choose a mix of 'anchors' (items that carry clear information about your invention, Hero, and Key Story) and 'curios' (items that seem intriguing without context, even if they aren't super relevant to the scenario or persona, as these out of place wonders help increase the illusion of a bigger world).

In *How To Future*, Scott Smith suggests drawing on what is familiar to your audience so that the familiar is juxtaposed with the unfamiliar, transporting the audience to the new world and challenging them to think and feel differently [1]. Refer to your earlier thoughts on your audience and choose what will resonate best with them.

Interaction mediums

When designing any interactive aspect of your prototype think in terms of tactile mediums and fun engagement. For example, make it possible for your audience to:

- Write or draw on paper or sticky notes to put on a wall
- Enter something into a computer
- Record a voice or video message (Consider privacy and consent when asking for or using these)
- Participate in a game
- Find something hidden

- Send an email

Creating echoes

Echoes are means of continuing the discussion beyond the sharing of your prototype.

Again, refer to your earlier thoughts on your audience to explore what will be the easiest ways for them to give feedback and/or share their experience of your prototype.

Echoes for your audience to share onward:

- Hashtags
- QR codes
- Collect emails to send them something as a take-away
- Suggest ways on how the audience can contribute to an issue identified in the process, like joining a related group or writing to an MP

Echoes for you, the designer, to revise after audiences have experienced the prototype:

- Collate the feedback/responses and share back on social media or via email, as a thank you
- Contribute yourself to an issue identified in the process, like joining a related group or writing to an MP
- Ask for feedback from experts—like scientists and engineers—or from potential users of the future product, then iterate and re-share
- Organise an ideation session with fellow designers, iterate, and re-share

Point of view

If you are using any kind of narration in your share back, consider who or what is sharing this future invention—the reporter/narrator doesn't have to be the same as the Hero or user:

- **1st person voice**—employs the I or we pronouns; spoken/written as from our own perspective and personal experience; "I opened the door to find this strange, hexagonal object suspended from the ceiling.

Steph looked scared."; the most immersive and most popular POV in literature.

- **2nd person voice**—you, your, and yours; "You open the door to find a strange, hexagonal object suspended from the ceiling. Steph looks scared."; not as popular as it requires a larger suspension of belief, and the different members of an audience may not relate to the words and tone.
- **3rd person limited**—he, she, them; a narration from a single character's perspective, and who can't know what other characters are thinking; "He opened the door to find a strange, hexagonal object suspended from the ceiling. Steph looked scared."
- **3rd person omniscient**—he, she, them; an all-knowing narration who can know what all characters are thinking; "He opened the door to find a strange, hexagonal object suspended from the ceiling. He thought that Steph looked scared, but she was pretending."
- **Multi-viewpoint**—a powerful way of communicating the complexity of a future

world from various perspectives without giving too much detail

Tone & language

While dystopian and utopian experiences can be more immediately impactful, keeping the tone more balanced will help the audience suspend their belief and immerse themselves deeper into the experience. Mix and match dystopian and utopian language in any words you use—refer to the marketing research on your **Research Map**. Consider what you learnt about your audience to use language and terminology they'll best respond to.

You can also balance tone with any decorative aspects such as colours, sound, etc. to bolster your preferred mood.

Understand your audience

- Understand their values to:
    - Know what they will most easily connect with
    - Be mindful of cultural sensitivities

- In what ways might they prefer to interact with your prototype—walking through an exhibition and taking photos, or trying some future food you cooked up?
- What would be the easiest ways for them to give feedback?

Leverage your strengths

Before you decide what form your prototype will take, consider what skills you (or your team) possess that you can leverage—can you edit video, create interactive prototypes, weld iron, etc.?

You could list your skills to match with the prototype forms below. But also push yourself to innovate on the forms you're not skilled in—if you're not a professional video editor, can you use your phone camera's in-built video editing to create a video?

And push yourself again to innovate in ways that don't cost a lot of money—see 'Prototype materials' below for ideas.

Get inspired

The best way to generate ideas for your playback is to look at how others have done speculative design. Grab another *Greedo's Lament*, sit back, and explore these wonders:

- postscapes.com/internet-of-things-award/design-fiction
- museumoffutures.com
- extrapolationfactory.com
- designedrealities.org
- designawards.core77.com

Pulling it together

The main things to remember about your prototype:

- It's a window into a world that generates wonder, not a solution
- It should possess mystery—As mentioned earlier, prototypes for speculative design are narrative-based, suggesting by their form and function the nature of their imagined

future world, while still leaving room for the audience to fill in the gaps. These personal interpretations encourage participants to articulate to themselves both their preferred future and their reasoning. Think of what would make sense in a normal conversation as a guide to how much detail to include.

- Use provocation—Include what you see as negative. Don't downplay what you might think is negative, treat it as a new norm
- If you're wanting to convey a message, avoid being too literal or preachy
- Be consistent with style

Physical prototype materials

Building something like a functioning future augmented reality contact lens wouldn't be practical for most of us, but often a mock-up with the right prototype form and tone will suffice. Here are a few affordable and practical materials and sources you can use:

- Papier-mâché
- Clay
- 3D printed objects

- Children's toys, medical and science kits, etc.
- A toy projector
- Second-hand stores
- Royalty-free images
- Royalty-free music and effects
- Stationary
- Craft supplies

Experiment

Remember, speculative design isn't just about experimenting with the future. It's about experimenting with the process, too. The *Future Scouting* method is one approach. Mix the tools and methods, invent your own approach, and have fun.

11
Home

Welcome back!

You made it home, back to your Now.

If you're still reading this—and sipping that green cocktail I promised you—you've well-earned candidacy to be a *Future Scout*.

You deserve a break. But I'm sure you'll soon be ready to explore the future again.

If you're looking for more future fun, here are some final tips:

- **Learn**—register your interest in learning how to practice Future Scouting at futurescouting.com.au
- **Play the Future Scouting game**—download the free, streamlined version of the book, made for playing with your team

- **Stay positive**—we all know the future won't be perfect, but if we can steer our Now with our hearts, values, and actions, as much as with our thoughts, we can turn passive wishing into 'active hope' [1]
- **Stay grounded**—playing in the future can get a little heady and leave us forgetful of today. Don't forget to focus some of your design superpowers on people and issues needing some attention now, including yourself, friends, and family, and community

One last thing

We need to talk about one more thing before a true *Future Scout* you become—the dangerous paradox of bringing something from the future to the present to stop it from existing tomorrow.

During one of your scoutings, you might stumble upon an advanced artefact that promises you riches and spoils at the expense of sharing knowledge. Tempting, it will be, to keep this something for yourself and benefit from its seductive future influences over the masses.

But *Future Scouts* have a code—we promise to honour the brave designers of the past by sharing all our pilfered bounties and to encourage alignment of the lifestyles our designs enable with sincerely defined values.

This is the way.

Happy hunting, *Future Scout*, and don't forget to keep the story going when you get back to the present any way you can.

Our preferred futures depend on it.

***

If you enjoyed this guide and resources, would you kindly give it a review? Reviews help self-funded books like *Future Scouting* get found.

And check out my other book,
The Life-centred Design Guide.

Thank you.

Appendix A
Prototype Examples

The following are examples of speculative design practitioners—artists, design studios, awards, and more— with prototype examples (sorted alphabetically).

Automato

Installation

This quirky design and research studio based in Shanghai combine the past of computer science, product design and electrical engineering to create designs and experiences on the fringes of product and interaction design.

automato.farm/portfolio/

automato.farm/experiments/

Beeple-crap

Artwork and illustration

And dropping a Speculative Design tab of acid takes you to beeple-crap on Instagram. If you haven't discovered this super-talented illustrator, animator, story-snippet-teller, then you better put the kids to bed and buckle up 'cause it ain't a family show. This is extreme, purely digital Speculative Design at its utmost, challenging viewers to evaluate popular culture and politics as one ongoing spectacle of outlandish techno-organic mutations.

Beeple can be found on Facebook, Instagram and Tik Tok.
instagram.com/beeple_crap/channel/?hl=en

CORE 77 Design awards

Various examples of award-winning designs

Projects, whether physically or digitally produced, designed for the purpose of cultural commentary, intervention, or exploration, or created as speculative design for a client or educational institution. Examples include future scenarios, social critique
designawards.core77.com/2020/speculative-design

Designed Realities Studio

Studio website with various examples

The Designed Realities Studio is a research and teaching platform led by Dunne and Raby at The New School in New York. Through practice-led research and project-based teaching, it aims to provide a rich and challenging context for students and faculty from disciplines across the university to integrate theory and practice into concrete responses to the complex fusion of politics and technology shaping today's social realities. designedrealities.org/

Euthanasia for everyone

Medical concept

The Speculative Design Provocative Award goes to *Soulaje*, a self-administered euthanasia wearable giving people control over the place and time of their death. Consisting of a vial of some death-inducing potion attached to a smartwatch, you simply tap and die. You can't get much more user-friendly than that! Created by Design Friction studio, All-Party Parliamentary Design and Innovation Group (APDIG), and Age UK, the *Soulaje* concept arose from exploring loneliness for the ageing. youtube.com/watch?v=tEyLzuULXKw

Extrapolation Factory

Real-world installations

Founded by creatives Chris Woebken and Elliott P. Montgomery, Extrapolation Factory use their own experimental, collaborative methods to explore democratised futures by creating hypothetical future props. They often embed their provocative props into real-world scenarios.

'99¢ Futures', the Factory's first project—and my personal favourite—began with a selection of possible future scenarios expanded into small stories. Related product ideas generated from the stories ('Benzene Vapor Refills, Mars Survival Kits, and Triple-Nipple Baby Bottles') were created via rapid-prototyping and then stocked for purchase in a real-world 99¢ store amongst its usual products. This guerrilla installation provoked conversations between strangers about the future scenarios triggered by these specials in the 99¢ store with Time-Warp sale.

extrapolationfactory.com/99-Futures

Institute for the future

Website with designs created for businesses

IFTF is a leading futures organisation. For over 50 years, businesses, governments, and social impact organisations have depended upon IFTF global forecasts, custom research, and foresight training to navigate complex change and develop world-ready strategies. IFTF methodologies and toolsets yield coherent views of transformative possibilities across all sectors that together support a more sustainable future. iftf.org/what-we-do/artifacts-from-the-future/

I wanna deliver a shark...

Medical and transhuman concept

Addressing the potential future issues of overpopulation and species extinction, Japanese designer Ai Hasegawa proposed future biomedical technology that enables women to use their reproductive system to birth endangered species... that they could also eat. Hasegawa likes to eat dolphin but recognises this adds to the animal's endangerment. She also wants to not waste her reproductive system but doesn't want children. Hasegawa took her concept further by conducting a scientific study to realise a scheme of how this could actually look.

aihasegawa.info/i-wanna-deliver-a-shark

grcd3021-f17.studiojunglecat.com/wp-content/uploads/2017/09/Ai-2.pdf

Museum of Futures

Virtual exhibitions of various examples

Not to be confused with The Museum of the Future under construction in Dubai, the Museum of Futures is an interactive virtual gallery created by Sydney-based creatives Claire Marshall and Mel Rumble. As a virtual visitor, you can tour the online gallery to explore artefacts from alternate futures created during experiential workshops facilitated by Marshall's creative company iflabs.

The latest exhibition, Pandemic Pivots, explores two possible futures — a utopia where we tackle climate change, and a dystopia where extreme climate change has altered our way of life forever. Previous exhibitions explored the Future of Work, the Future of Australia,

and The Future of Food. The virtual Museum of Futures showcases the fascinating artefacts from all exhibitions, including a 'repurposable' toy made in 2035, a sculpture commemorating the 2030 introduction of The Native Species Title Act that 'gave animals rights to all native forests and bushlands across Australia', and a bottle of Australian air from the early 2020s'.
museumoffutures.com

Nefula

Exhibition, interactive

Nefula is the first Italian studio of Near Future Design
nefula.com/work/

Neo Fruit

Food concept

Artist and designer Ai Hasegawa explored advanced genetic engineering and the potential future food shortage and proposed the 'Pop Roach'—a genetically modified roach redesigned into a 'cute, colourful, tasty, and high nutritional' version of the less appealing black and brown versions. While we're on the topic of food, design student Meydan Levy received recognition in the

Student Winner 2019 Core77 Speculative Design Awards for her Neo Fruit, a vision of artificially created fruits suggested as future food consumption in an overpopulated world.
designawards.core77.com/speculative-design/84449/Neo-Fruit

Dunne & Raby

Various examples

Anthony Dunne and Fiona Raby's award-winning design studio uses design as a medium to stimulate discussion and debate amongst designers, industry, and the public about the social, cultural, and ethical implications of existing and emerging technologies.
dunneandraby.co.uk/content/projects

Their groundbreaking book, Speculative Everything, is the definitive read on speculative design and also features various examples.
Speculative Everything

SPUTNIKO!

Film and multimedia

Born in 1985, Sputniko! is a Japanese/British artist based in Tokyo. Sputniko! is known for her film and multi-media installation works which explore the social and ethical implications of emerging technologies. sputniko.com/Works-1

Threatcasting Lab

Graphic novels and motion comics

The mission of the Threatcasting Lab at Arizona State University is to serve as the premier resource for strategic insight, teaching materials, and exceptional subject matter expertise on Threatcasting, envisioning possible threats ten years in the future. threatcasting.asu.edu/scifi-prototypes

Appendix B
Notes

Chapter 1
[1] Kelly, Kevin. Protopia. kk.org/thetechnium/protopia , 2011.
Chapter 2
[1] Dunne, Anthony; Raby, Fiona. *Speculative Everything.* MIT Press, 2013, U.S.A. Page 2.
[2] Textor, Robert B. *A Handbook on Ethnographic Futures Research.* Stanford University, 1980, U.S.A. Page 13.
[3] Hancock, Trevor; Bezold, Clement. *Possible futures, preferable futures.* An Overview of the Health Futures Field. WHO, 1993, Geneva. Page 11.
[4] Mescia, Adriano. *Design Thinking and Speculative Design*, adrianomescia.com/design-thinking-and-speculative-design , 2015.
[5] Masaki Iwabuchi. *Design for the 22nd Century*, masakiiwabuchi.me , 2020.
[6] X, The Moonshot Factory. *Loon*, x.company/projects/loon , 2018.
Chapter 3
[1] Artefact Group. *The Tarot Cards of Tech*, tarotcardsoftech.artefactgroup.com , 2017.
[2] Board Of Innovation, *Future Scan*, boardofinnovation.com/tools/future-scan , 2020.
[3] Fox, William James. *Future Timeline*, futuretimeline.net , 2018.
[4] Gray, Dave. *Cover Story*, gamestorming.com/cover-story , 2010.
[5] Sibbet, David. *Cover Story Vision® Canvas*, designabetterbusiness.tools/tools/cover-story-canvas , 2019.
[6] Candy, Stuart. *The Thing From The Future*, situationlab.org/project/the-thing-from-the-future , 2015.

[7] Wilson, Mark. *Frog's Five Steps to Predicting The Future*, fastcompany.com/3065080/frogs-5-steps-to-predicting-the-future , 2016.
[8] Candy, Stuart; Kornet, Kelly. *A Field Guide to Ethnographic Experiential Futures*, researchgate.net/publication/317837102_A_Field_Guide_to_Ethnographic_Experiential_Futures , 2017.
[9] Johnson, Brian David. *Science Fiction Prototyping: Designing the Future with Science Fiction*

Chapter 4
[1] Dunne, Anthony; Raby, Fiona, *Speculative Everything*, MIT Press, 2013, U.S.A. Page 2.

Chapter 5
[1] Candy, Stuart; Kornet, Kelly. *A Field Guide to Ethnographic Experiential Futures*. researchgate.net/publication/317837102_A_Field_Guide_to_Ethnographic_Experiential_Futures , 2017.
[2] Hershfield, Hal. *Future self-continuity: How conceptions of the future self transform intertemporal choice* ,researchgate.net/publication/51738840_Future_self-continuity_How_conceptions_of_the_future_self_transform_intertemporal_choice , Annals of the New York Academy of Sciences, 2011.
[3] Bryan, Dr. Tomi White. *Personal Core Values: The Definitive Guide (Worksheet + Flashcards)*, tomillama.com/personal-core-values , 2020.

Chapter 6
[1] Smith, Scott; Ashby, Madeline. *How To Future*, Kogan Page Limited, U.S.A. 2020. Page 69-70.
[2] Institute For The Future. *Searching for Signals: Tips & Tricks from IFTF Staff*, iftf.org/future-now/article-detail/searching-for-signals-tips-trick-from-iftf-staff , 2020.
[3] Smith, Scott; Ashby, Madeline. *How To Future*, Kogan Page Limited, 2020. U.S.A.
[4] Zaidi, Leah. Building Brave New Worlds: Science Fiction and Transition Design,

researchgate.net/publication/321886159_Building_Brave_New_Worlds_Science_Fiction_and_Transition_Design , 2017.

Chapter 7

[1] Gray, Dave. *Squiggle Birds,* gamestorming.com/squiggle-birds, Gamestorming, 2015.

[2] Abrams, Mere. *64 Terms That Describe Gender Identity and Expression,* healthline.com/health/different-genders , Healthline, 2019.

[3] Foley, R. A.; Lahr, M. Mirazón. *The evolution of the diversity of cultures,* ncbi.nlm.nih.gov/pmc/articles/PMC3049104, 2011.

[4] Australian National University. *Different types of disabilities,* services.anu.edu.au/human-resources/respect-inclusion/different-types-of-disabilities .

[5] Shaw, Alison. *Designing for inclusivity: How and why to get started,* invisionapp.com/inside-design/designing-for-inclusivity , Inside Design, 2018.

Chapter 8

[1] Glenn, Jerome C. *The Futures Wheel - The Millennium Project,* millennium-project.org/wp-content/uploads/2020/02/06-Futures-Wheel.pdf , 1971.

Chapter 10

[1] Smith, Scott; Ashby, Madeline. *How To Future,* Kogan Page Limited, U.S.A. 2020. Page 151.

[2] Harrison, M John. *If the aliens lay eggs, how does that affect architecture?,* The Guardian, theguardian.com/books/2021/jan/05/if-the-aliens-lay-eggs-how-does-that-affect-architecture-sci-fi-writers-on-how-they-build-their-worlds, 2020.

Chapter 11

[1] Macy, Joanna; Johnstone, Chris. *Active Hope,* activehope.info , 2012.

About the author

Damien Lutz is a UX Designer, researcher, self-published sci-fi author, and author of *The Life-centred Design Guide*, a practical exploration of how human-centred design is expanding to consider all peoples, animals, and planet.

futurescouting.com.au
medium.com/@damienlutz
twitter.com/_the_future
instagram.com/future.scouting

www.ingramcontent.com/pod-product-compliance
Lightning Source LLC
Chambersburg PA
CBHW020256030426
42336CB00010B/795